WORKING T PROTEC...
THE PLANET

Cynthia O'Brien

CRABTREE
PUBLISHING COMPANY
WWW.CRABTREEBOOKS.COM

CRABTREE
PUBLISHING COMPANY
WWW.CRABTREEBOOKS.COM

Author: Cynthia O'Brien
Picture Manager: Sophie Mortimer
Designer: Lynne Lennon
Design Manager: Keith Davis
Children's Publisher: Anne O'Daly
Editorial director: Kathy Middleton
Editor: Janine Deschenes
Proofreader: Wendy Scavuzzo
Print coordinator: Katherine Berti

Copyright © Brown Bear
Books Ltd 2020

Photographs: (t=top, b= bottom, l=left, r=right, c=center)

Front Cover: Shutterstock: Diego G Diaz (center right), FooTToo (bottom left), Traci Hahn (bottom center); Wikimedia Commons, Division of the Ministry of Information and Broadcasting (center left)

Interior: Alamy: ClassicStock 25, Dylan News Images 32, Zuma Press Inc 35; Dreamstime; Shawn Goldberg 36; Getty Images: Eric Baradat 41t, Bettmann 23b, CBS Photo Archive 21, Vernon Meritt II/The LIFE Picture Collection 22; Micheline Pelletier 27; iStock: Bakstad 16, BeyondImages 8, Igorokov 20, Willia Sherman 26; Library of Congress: 17; NASA: Rye Livingston 34; Public Domain: Kalpit Bishnol 12, Miranda Productions Inc/Miranda Smith 31, Robarts- University of Toronto 14, USGS/Lovecz 33, Uyvsdi 15, Yale University Libraries 18; Shutterstock: Africa Studio 23t, Arindam Banerjee 29, Sheila Fitzgerald 7, Rich Carey 6, ChameleionsEye 24, Daniele Cossu 37, 44, FloridaStock 5b, FooTToo 1, 5t, John Gomez 41b, Holli 9, Ink Drop 43, Sergey Panikratov 11, Photo Image 28, ramji Creations 4, Iryna Rasko 13, Andrey Shcherbukhin 10, Rachel Warriner 40; Tactile Tends: Caroline Power 39; USDA: Pacific Southwest Forest Service 19; Wikimedia Commons: P199 30.

Brown Bear Books has made every attempt to contact the copyright holder. If you have any information about omissions, please contact licensing@brownbearbooks.co.uk

Library and Archives Canada Cataloging in Publication

Title: Working toward protecting the planet / Cynthia O'Brien.
Names: O'Brien, Cynthia (Cynthia J.), author.
Description: Series statement: Achieving social change | Includes bibliographical references and index.
Identifiers: Canadiana (print) 20200299247 | Canadiana (ebook) 20200299298 | ISBN 9780778779520 (softcover) | ISBN 9780778779469 (hardcover) | ISBN 9781427125507 (HTML)
Subjects: LCSH: Environmentalism—Juvenile literature. | LCSH: Environmental protection—Citizen participation—Juvenile literature. | LCSH: Green movement—Juvenile literature.
Classification: LCC GE195.5 .O27 2021 | DDC j333.72—dc23

Library of Congress Cataloging-in-Publication Data

Names: O'Brien, Cynthia (Cynthia J.) author.
Title: Working toward protecting the planet / Cynthia O'Brien.
Description: New York : Crabtree Publishing Company, 2021. | Series: Achieving social change | Includes bibliographical references and index.
Identifiers: LCCN 2020032434 (print) | LCCN 2020032435 (ebook) | ISBN 9780778779469 (hardcover) | ISBN 9780778779520 (paperback) | ISBN 9781427125507 (ebook)
Subjects: LCSH: Environmentalism--Juvenile literature. | Environmental protection--Citizen participation--Juvenile literature. | Green movement--Juvenile literature.
Classification: LCC GE195.5 .O27 2021 (print) | LCC GE195.5 (ebook) | DDC 363.7--dc23
LC record available at https://lccn.loc.gov/2020032434
LC ebook record available at https://lccn.loc.gov/2020032435

Crabtree Publishing Company
www.crabtreebooks.com   1-800-387-7650

Published in Canada
Crabtree Publishing
616 Welland Ave.
St. Catharines, ON
L2M 5V6

Published in the United States
Crabtree Publishing
347 Fifth Ave
Suite 1402-145
New York, NY 10016

Published by CRABTREE PUBLISHING COMPANY in 2021

Printed in the U.S.A./092020/CG20200810

CONTENTS

INTRODUCTION

Protecting the planet means protecting Earth's natural environment, including its atmosphere, land, oceans, and climate. Activists are the people who work to create a sustainable planet for generations to come.

Activism can take many forms. Some people write books that inspire others to speak up. Others write letters to the government or start petitions. Activists might lead demonstrations and give speeches to groups of people to make their message clear. Activist groups can force governments to listen and to make changes to laws. Their organizations help to make these changes happen. They can also advocate for people experiencing injustices.

*This activist has chained herself to a **logging** machine to protest cutting down a forest in Australia. Her **tactic** forces others to pay attention to her cause.*

Activists organize marches to raise public awareness about issues affecting the environment.

Protecting the Planet

Many problems that endanger the planet are caused by humans, such as clear-cutting forests to make way for agriculture, and burning **fossil fuels** which causes **global warming**. Humans also cause a lot of **pollution**—from plastic that chokes ocean wildlife to chemical pollution that threatens habitats. Environmental activists have been working for many years to try to change the way that people treat Earth. Today, people are much more aware of the need to be careful with Earth's resources. Research, such as opinion polls, shows that many see climate change as a priority. There is still a long way to go, and work for activists to do.

One of the effects of global warming is the melting of sea ice in the Arctic, which threatens the polar bears that live there. The animals use the ice to hunt and are at risk of starvation.

CHAPTER 1

WHY PROTECT THE PLANET?

People started farming around 15,000 years ago. As farms grew larger, people cut down forests and drained swampland. Today, people use about half of all habitable land for agriculture.

Indigenous peoples grew crops native to where they lived, such as squash, corn, and beans. These plants provided a healthy diet for people, but also left behind nutrients in the soil. Later farming practices, such as cutting down large areas of forest around the world, eroded the soil. In the mid-1900s, farmers started using chemical pesticides to manage their crops. These chemicals leaked into the soil and ran off into water sources.

This is a palm oil plantation in Indonesia. Rainforest trees are being cut down to clear the land for palm oil trees. This damages soil and destroys natural habitats.

Given the environmental effects of raising livestock, activist groups such as Greenpeace suggest that people should reduce their meat consumption.

Farming and the Environment

Another huge issue for the planet is raising livestock. Farmers use about 26 percent of the world's land for grazing. They also use one-third of all **arable** land to grow food for livestock. This practice has led to water and air pollution. Livestock produce methane, which is a **greenhouse gas**. Clearing the land for farms, cities, and **industry** has caused widespread **deforestation** around the world.

Key Voices

Rainforest Action Network

Since 2013, Rainforest Action Network has targeted 20 major snack food producers, which are major users of palm oil. Their campaign pushes snack food companies to switch to using sustainable palm oil that has not been produced by companies that cut down rain forests or exploit local communities and workers.

Depending on Resources

As societies developed, people used fuels for industry and minerals to make machines. Earth supplied these natural resources. **Most people thought that these supplies would never run out.**

Mining is a process that involves getting minerals or other substances from the ground. People dig into the ground to **extract** gold, silver, coal, diamonds, and more. All mining disturbs the environment. Some of the more harmful effects are eroding the soil and loss of **biodiversity**. Chemicals used in mining leak into surrounding water sources and pollute them. Air pollution is another problem since mining can release **toxic** chemicals into the air.

Mining removes trees and topsoil, destroying the landscape and wildlife habitats.

Key Voices

Earthworks

Earthworks is a non-profit organization (NPO) that uses scientific findings to expose the health and environmental impact of mining. Formed in 2005, the organization works with local communities to take action against oil and gas companies that damage the environment. To achieve change, Earthworks uses tactics such as petitions, protests, lobbying politicians, and legal actions.

Drilling for Oil and Gas

Oil and gas also come from the ground. People have come to depend on these substances for fuel and for making products such as plastic. Oil and gas are **non-renewable** resources. To access oil and gas, drills go deep underground, or underwater into the ocean floor. This can disturb habitats and cause dangerous spills. Water and air pollution are problems, too.

Burning oil and gas releases greenhouse gases into the atmosphere. Over time, these gases have caused the planet to heat up and led to **climate change**. Environmental activists are working to stop global warming before it gets any worse. They raise awareness about the problem and push governments to make changes that reduce greenhouse gas emissions.

*These Aboriginal people in Australia protest against **fracking** for oil and gas. Fracking causes air and water pollution, and can trigger earthquakes.*

Counting the Cost

Human activities have damaged ecosystems and polluted oceans, air, and soil. They have also led to climate change, causing the ice caps to melt, sea levels to rise, and extreme weather such as hurricanes. But humans need a healthy planet to survive.

Humans need clean, fresh water and healthy soil in which to grow crops. They need clean air to breathe and homes that are safe from flooding. Plants and animals should be free from **contamination**. This is not only so that humans can eat them. All living things on Earth are part of ecosystems. They live in all kinds of environments to keep the planet in balance.

Orangutans are under threat. Their rainforest habitats are being cut down for logging, mining, and to clear space for farmland.

Activists work in many ways. These World Wildlife Fund (WWF) supporters are organizing a craft fair to raise money for the charity.

Different Issues

Protecting the planet is an ongoing job. Environmental activists have been on the front line for hundreds of years. From protecting forests from logging to raising awareness about global warming, many have made huge changes or tried to cause change. There have been different issues to tackle, but their basic message is the same. People must respect and protect Earth, its water, air, and soil, and all its creatures.

Key Voices

WWF

The World Wildlife Fund (WWF) started in 1961. It now works in over 100 countries around the world. WWF protects endangered species by carrying out wildlife surveys and setting up anti-poaching projects. One of the current campaigns is TX2, which aims to double the number of wild tigers between 2010 and 2022.

CHAPTER 2

GREEN BEGINNINGS

In 1730, the king of Jodhpur, India, sent his army to collect timber to build his palace. Large forests of khejri trees grew nearby. Local villagers held the trees sacred. They were determined to protect them.

The villagers belonged to the Bishnoi sect of the Hindu religion. Founded in 1485 in India, its followers believe that all nature is sacred and that no trees should be cut down. When the villagers learned about the king's plan, they knew they had to do something. One of the village women, Amrita Devi, threw her arms around a khejri tree. Her three daughters copied her, each hugging a tree. The women refused to move.

Each year the Bishnoi community commemorates Amrita Devi and the other villagers who were killed saving the trees.

> If a tree is saved even at the cost of one's head, it is worth it.
>
> Amrita Devi

Khejri trees are an important part of the ecosystem in the dry Thar Desert of northwestern India.

Tree Huggers

The soldiers killed Amrita and her daughters and started cutting down trees. They warned the other Bishnoi people not to protest or they would die, too. But this warning did not stop the villagers. People of all ages walked to the forest to hug the trees in protest. The army killed 363 people.

When the king heard about the villagers' courage, he told his army to stop. He passed a law to protect the khejri forests. He banned people from cutting down trees and also stopped all hunting in the area. Today, a memorial to the Bishnoi tree huggers stands in Khejarli village. The term "tree hugger" is still used for environmental activists.

Early Environmentalism

Protecting the planet has a long history. Ancient Greek and Roman thinkers wrote about the harmful effects agriculture had on land, for example. But most people did not connect human activities with damage to Earth.

As cities grew, air pollution became a problem. Until the 12th century, people in London used wood as fuel. As the city's population grew in the 1300s, and forests were cut down, people switched to burning coal. **Sea coal** was cheap and plentiful but it gave off a lot of smoke. This created thick **smog** in London. The situation became so bad that in 1306, King Edward I banned coal burning in the city. However, faced with a need for fuel, and with wood being less available, people continued to burn coal.

By the 1870s, factories relied on coal to make their goods. Smoke from poorly designed chimneys polluted the air.

Caretakers

As Europeans settled in other continents, they cleared lands to build cities and roads. They worked the land to create large farms, and built dams to redirect natural river systems. The new settlers ignored how Indigenous peoples used land in sustainable ways. They had been farming the land for hundreds of years. They rotated crops to keep the soil healthy, and used food waste to fertilize the land. Many Indigenous nations are still fighting for the right to protect the lands on which their ancestors lived.

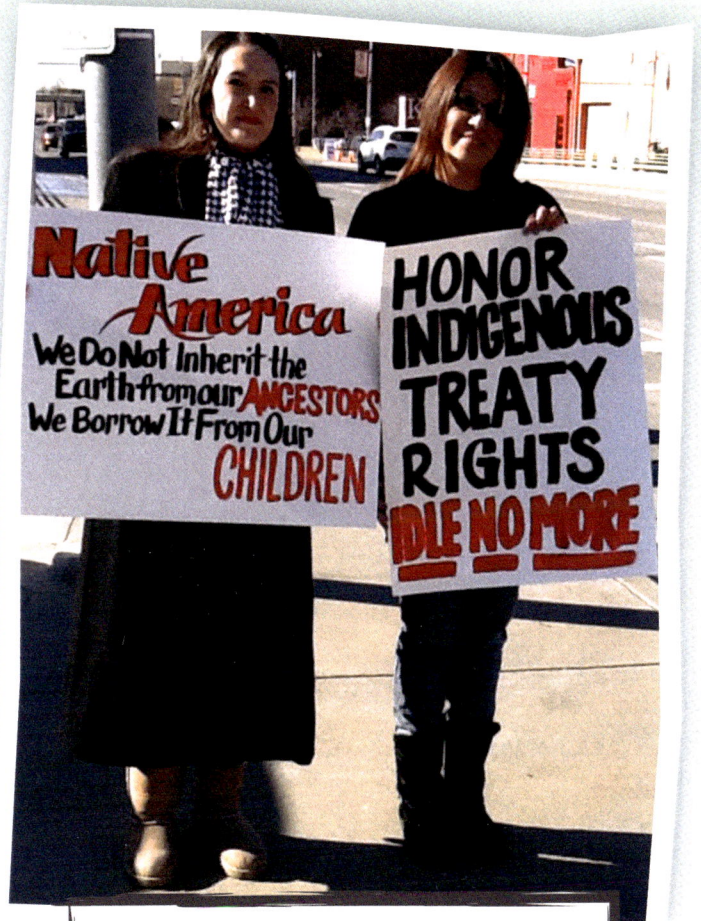

Indigenous peoples are still leading voices in the fight to save the planet. The Idle No More movement, set up by Indigenous peoples in Canada, stages peaceful protests to push for better environmental protections.

Key Events

Laws for Clean Air

Edward I's ruling was the first clean air law. By the 20th century, air pollution's impact was clear. In London, England, smog killed up to 12,000 people in 1952. That led to the Clean Air Act of 1956, which set up smokeless zones. It was a key step in the development of laws to protect the environment. The U.S. Air Pollution Control Act of 1955 was the first in the United States to deal with air quality at a national level.

CHAPTER 3

A MOVEMENT BEGINS

John Muir moved with his family from Scotland to the United States in 1849. After working and studying, Muir took a long trip, walking and sailing. He sailed to Cuba and Panama before making his way up the California coast.

Muir was awestruck by California's wilderness. He was especially drawn to the Sierra Nevada mountains and Yosemite. Muir built a cabin in the Yosemite Valley. He began writing about nature and **conservation**. Meanwhile, Muir continued to travel. He visited Alaska, South America, Australia, Africa, China, and Japan. When he returned to California, he noticed the damage that cattle and sheep farming had caused to the fields and forests.

Tenaya Lake in Yosemite National Park is surrounded by forests and is home to black bears. Muir was determined to protect the environment there.

> In every walk
> with nature one
> receives far more
> than he seeks.
> John Muir

In 1903, Muir (right) took President Theodore Roosevelt on a three-day camping trip in Yosemite National Park. He persuaded the president to make Yosemite Valley and Mariposa Grove part of Yosemite National Park.

A National Park

Muir wanted to protect California's wilderness from all farming, settlement, and industry. He wrote essays in magazines and letters to the government. This activism helped to create Yosemite National Park in 1890. Two years later, Muir was one of the founders of the Sierra Club. The organization set out to preserve the Sierra Nevada and other natural areas. Muir helped to establish other national parks, such as the Grand Canyon. Muir died in 1914, but the Sierra Club continued to grow and is still active today. The government created more national parks and established the National Park Service, which protects and conserves the country's natural places.

Worries and Warnings

The environmental movement started to take shape in the 1800s. In many ways, it was a reaction to the Industrial Revolution, which changed the way people lived and worked.

By the middle of the 1800s, steel and coal were big industries. In the United States and Canada, governments forced Indigenous peoples from their traditional lands. White settlers took over, farming and building towns. Railway lines, roads, and cities spread over the landscape. This destroyed habitats and also led to deforestation. Factories increased air pollution and let waste run into nearby rivers and streams.

*The first **transcontinental** railroad in the United States was finished in 1869. Canada's first transcontinental railway was completed in 1886.*

A Spark for a Movement

These issues sparked a conservation movement to conserve natural resources such as wood and water. In his 1864 book, *Man and Nature*, American conservationist George Perkins Marsh pointed out the damage that human actions were causing to the environment. He saw that deforestation led to **desertification**. He urged people to conserve Earth's resources. In Canada, loggers were some of the first people to be concerned about the environment. The Canadian Forestry Association was founded in 1900 to promote the conservation of forests.

Meanwhile, scientists were making discoveries that are still important today. In 1856, an American scientist, Eunice Newton Foote, proposed that carbon dioxide in the **atmosphere** would make Earth warmer. She published an article on her findings, probably the first warning about global warming.

Key Voices

Aldo Leopold

Aldo Leopold was influenced by the ideas of John Muir. Leopold was born in Iowa in 1887. Leopold promoted the establishment of official wilderness areas. The first of these was Gila National Forest, which was established in 1924. Leopold's book, *A Sand County Almanac*, urged people to preserve ecosystems and to protect the natural world.

Harmful Chemicals

The fight to preserve the wilderness continued into the 1900s. Activism led to governments establishing national parks and wilderness areas. At the same time, the agriculture industry grew, and other issues were brewing.

Scientists developed pesticides, which are chemicals that are sprayed on crops to kill pests. One of the most widely used was DDT. Farmers used it on food crops in the United States, Canada, and across the world. DDT was effective and cheap to produce. Rain did not wash it away easily. It protected crops from insects such as the Colorado potato beetle. However, other animals that ate the insects also became infected.

Fish absorbed DDT that ran into the waterways. Bald eagles were poisoned when they ate the contaminated fish.

Key Voices

Environmental Defense Fund

The Environmental Defense Fund (EDF) is a **grassroots** movement that brings together scientists and lawyers. It won a federal ban on DDT in the United States in 1972 and has gone on to fight for other environmental protections.

A Writer Who Dared

In 1962, Rachel Carson dared to point out how dangerous pesticides are to animals, air, soil, and humans. Carson was an American marine biologist and a talented writer. She became increasingly concerned about the use of chemical pesticides. The result was her important book, *Silent Spring*. Carson wrote that powerful chemical pesticides, such as DDT, were poisoning entire **food chains**.

Carson's book challenged the agriculture and chemical industries, and they tried to **discredit** her work. However, the book came to the attention of President John F. Kennedy. Carson spoke to the U.S. Senate in 1963 to call for stronger pollution laws. The use of DDT came under stricter supervision. Carson's work was the start of a movement to ban DDT. However, it wasn't until 1972 that DDT was banned for agricultural use in the United States and Canada.

In 1963, Carson spoke about her findings in a national TV report, "The Silent Spring of Rachel Carson." Her calm authority won over many viewers and the program made pesticide use a major public issue.

CHAPTER 4

GREEN POLITICS

On the morning of January 28, 1969, oil began to gush into the ocean. The Union Oil Company was drilling into the ocean floor. A buildup of pressure had caused oil and gas to burst through cracks in the ocean floor.

The oil company worked desperately to contain the spill. By the time they stopped the leaks, about 3 million gallons (1.4 million l) of oil had spread across the water. A black slick covered 35 miles (56 km) of water along the coast. The eruption covered sea birds with oil and poisoned marine life. The oil spill killed between 3,700 and 9,000 birds.

This aerial picture shows the oil leaking from the Union Oil Company platform and spreading across the ocean.

From Disaster to Action

The Santa Barbara oil spill was the largest in American history at the time, and sparked a fierce reaction. Local environmental organizations, such as Get Oil Out! (GOO!) and the Community Environmental Council, formed quickly. They called on the government to act on environmental issues. GOO! collected 100,000 signatures on a petition to ban offshore drilling. The protests led to a surge in political action. President Richard Nixon signed the National Environmental Policy Act of 1969. This was followed a year later by the Environmental Protection Agency. The disaster inspired a generation of activists.

I just instantly thought, this is going to change the world.
Paul Relis, a founder of the Community Environmental Council

Local people rushed to the beaches to help. They raked straw along the shore line to try to soak up the oil.

Green Activism

Rachel Carson's book and the Santa Barbara oil spill showed how modern technologies could damage the environment. New groups were set up to press governments to take action.

Greenpeace was founded by activists in Canada in 1971. For its first action, a small group sailed from Vancouver to Alaska. They were protesting **nuclear testing**. Their action brought worldwide attention to the risks of nuclear testing. Greenpeace soon expanded its work to include other issues. Today, its members sign petitions, write letters to politicians, and take part in marches.

Greenpeace uses ships to engage in peaceful protests, including confronting illegal fishing boats or getting between whale hunting ships and their targets. The ships also draw attention to different causes.

These people protested peacefully on the first Earth Day. Today, Earth Day activism includes asking supporters to record air quality and plastic pollution with a special app.

Earth Day

People came together on April 22, 1970. This was the first Earth Day. About 20 million people took part across the United States. They held demonstrations, speeches, meetings, and teach-ins. People spoke and learned about environmental issues. They focused on government action on these issues. In 1990, Earth Day became a worldwide event. Every year, more than 1 billion people in more than 190 countries recognize the day.

Key Events

Green Political Parties

The first Green political parties started in 1972 in Tasmania and New Zealand. Green Party politician Caroline Lucas was elected as a government representative in the United Kingdom in 2010. Canada's Green Party was established in 1983. Green Party leader Elizabeth May was elected as a government representative in Canada in 2011. Support for Green parties is increasing in Canada, with three more representatives being elected in 2019.

Making Changes

Events such as Earth Day pushed governments to take action. Two American government agencies were formed in 1970. The Canadian government created Environment Canada in 1971.

The National Oceanic and Atmospheric Administration (NOAA), formed in 1970, oversees climate changes and works to protect ecosystems. By the end of the year, the Environmental Protection Agency was also in place. Canada started with the Arctic Waters Pollution Prevention Act in 1970. After decades of work by activists, these governments were now actively involved in protecting the planet.

The NOAA's research ship Fairweather *is used to map the oceans and to collect scientific samples from the ocean depths.*

International Movements

Environmental activists in other parts of the world also took action and put pressure on governments. In Kenya, activist Wangari Maathai created the Green Belt Movement in 1977. She protested against the destruction of forests to make way for farmland. The Chipko movement in India also fought the government to protect trees in efforts similar to the 1730 activists who came before them. The mostly female activists used peaceful tactics including hugging trees and confiscating logging machinery. The first tree-hugging demonstrations led other communities to adopt the same methods. The protests led to government action against deforestation.

Key Events

Wangari Maathai

Kenyan scientist, Wangari Maathai, was active in the National Council of Women in Kenya when she introduced the idea of community tree-planting. Her Green Belt Movement, created in 1977, tackled poverty and damage to the environment. It also involved community tree-planting projects that helped restore the land and plant more than 51 million trees. Maathai won the 2004 Nobel Peace Prize for her work.

CHAPTER 5

PLANET IN CRISIS

In 2016, Dakota and Lakota people of the Standing Rock Sioux Tribe came together to protest the Dakota Access Pipeline. The pipeline was being built to carry oil from the Bakken oilfield in North Dakota to a refinery in Patoka, Illinois, near Chicago.

Part of the pipeline would travel through the Standing Rock Reservation and cross rivers, including the Missouri. The Missouri is the main freshwater supply for the reservation, and people feared that the pipeline could contaminate drinking water. The pipeline's course would also take it close to places that were important to the Standing Rock Sioux people, including sacred, or holy, burial sites.

Rallies to support the protestors took place around North America. Like this protest in Toronto, many were led by Indigenous peoples.

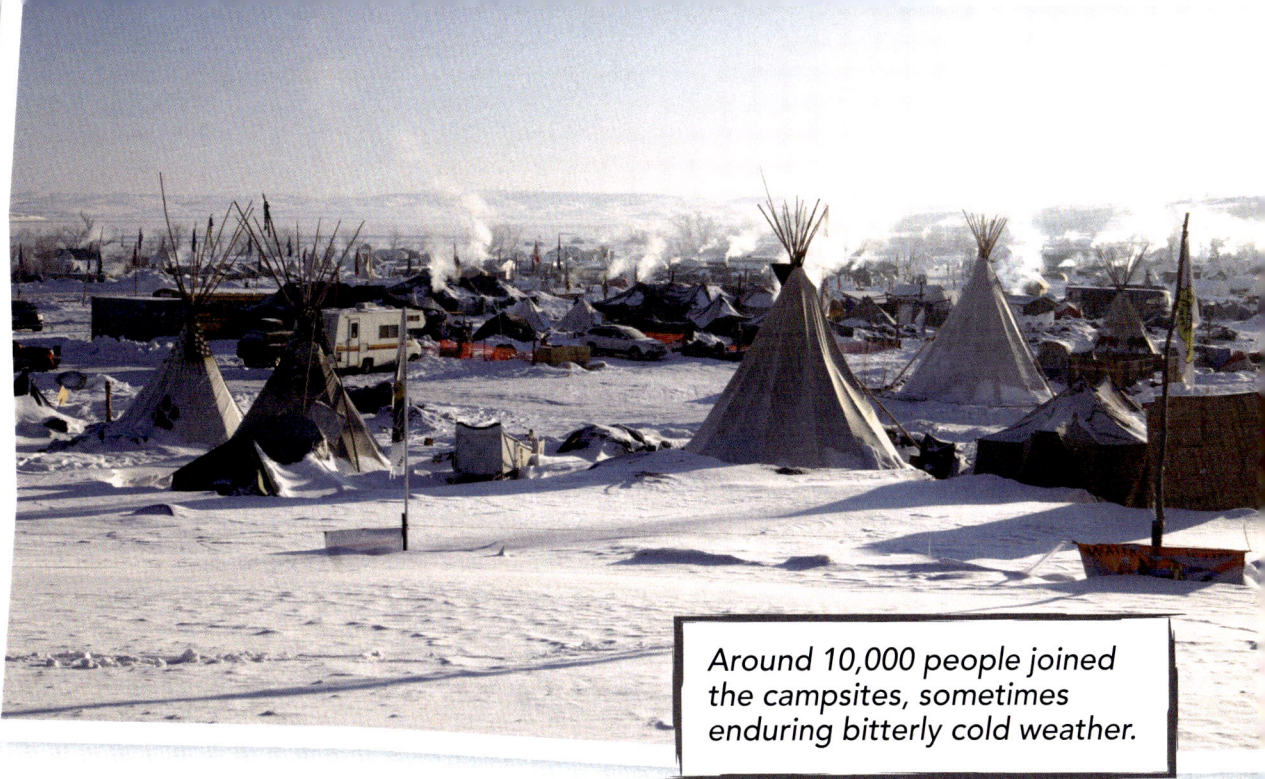

Around 10,000 people joined the campsites, sometimes enduring bitterly cold weather.

Protest Camps and Legal Action

The Standing Rock Sioux people led the protest. Members of the tribe set up a camp in April 2016 to try to block the construction work. They were joined by thousands of Indigenous peoples and climate activists, and more camps sprang up. The activists called themselves "water protectors." The Standing Rock Sioux people complained that the project had gone ahead without a proper study of its environmental impact, and without consulting tribal leaders.

The protest was covered by media around the world and gained worldwide support. The United Nations backed the calls to halt construction of the pipeline. It also condemned local security forces for using excess violence. In 2016, construction was stopped temporarily to allow for an environmental impact survey. When President Trump was elected, he supported the pipeline and it started transporting oil in 2017. However, in 2020, the Standing Rock activists won an important court action. A judge ruled that the pipeline must temporarily stop operating and empty its oil so a full environmental impact report can be made.

Earth Protectors

Indigenous activists are strong voices against the building of oil pipelines. They are part of a long history of Indigenous peoples around the world who are active in protecting Earth.

In 1990, the Indigenous Environmental Network (IEN) formed in the United States. Each year, the group holds a conference called the Protecting Mother Earth Gathering, which brings Indigenous activists together. The IEN campaigns to protect lands from toxic waste, mining activities, deforestation, and other environmental harm. In 1992, the U.S. Environmental Protection Agency developed a justice program to work with tribes to develop environmental campaigns. In the same year, Canada passed the Convention on Biological Diversity as a step toward giving Indigenous peoples a more active role in conserving ecosystems and managing resources.

Starting in 1971, Cree activists in Quebec tried to block the construction of a new hydroelectric dam system. Though some construction went forward, the Quebec government agreed to give the Cree more land rights in exchange for their agreement that the project be finished.

Campaigns for Many Causes

Meanwhile, activists campaigned for other causes such as repairing the hole in the **ozone** layer, and saving the rain forests. The Montreal Protocol in 1987 was agreed on by every country on Earth. They agreed to ban the chemicals that had caused the hole. Today, much of the ozone layer has been restored.

Other movements were slower, but still effective. The Rainforest Action Network set up a **boycott** of Burger King restaurants because they used beef from Costa Rica. Ranchers in Costa Rica had destroyed rain forest to breed the cattle. The plan worked, and Burger King stopped using the beef. By 1998, the activists had stopped deforestation in Costa Rica.

Thanks to Mendes, large areas of the Amazon rain forest are protected for Indigenous peoples. However, deforestation is still a concern there.

Key Voices

Chico Mendes

Chico Mendes collected rubber in the Amazon rain forest in Brazil. He and others were thrown out of their jobs to make way for cattle ranches. Mendes fought to protect the rain forest from deforestation. He called for forest reserves, or protected areas, to be set up so that products could be collected sustainably. This was important for workers and Indigenous peoples who depended on these products. He persuaded the government to create the first reserve in 1988. Mendes was killed by angry cattle ranchers later that year.

Backward and Forward

In the 1980s, governments and industries, especially in the United States, pushed back against the ideas of the environmental activists. This reversed some of the successes of the 1970s, but sparked more activism, as well.

The Sagebrush Rebellion in the late 1970s involved local politicians, miners, and ranchers in the western United States. Much of the land was owned by the federal government. The protestors wanted the land to be under local control. They disliked the strict environmental laws passed in the 1970s, which they thought affected their businesses. In 1981, Ronald Reagan became president of the United States. He pulled back government support of environmental causes.

This Earth First! activist is being arrested at a protest in London, England. Activists are often arrested while fighting for their cause.

Dead and dying trees are a sign of acid rain. The rain strips the soil of nutrients that the trees need to grow.

New Issues, New Action

Activists did not give up. Earth First!, an organization founded in the southwestern United States, arranged more direct protests. Their members blocked the way of bulldozers that were going to clear a forest. Other grassroots groups formed to focus on local issues. In Warren County, North Carolina, a group of mostly Black activists staged a peaceful sit-in protest against the construction of a landfill for toxic chemical waste. A number of campaigners were arrested, and the construction went ahead.

Another problem was **acid rain**. This is made when industries release chemicals that mix with air high up in the atmosphere. The chemicals become part of rain and snow and fall back to the ground. Acid rain kills trees and poisons waterways. By 1990, environmentalists pushed governments to do something. The U.S. government updated the Clean Air Act of 1970. It added the Acid Rain Program, which limits the amount of chemicals that power plants can release. In 1994, Canada and 25 other countries signed a United Nations agreement to reduce emissions of sulfur, a major cause of acid rain.

Raising the Alarm

Global warming and climate change were not huge topics before the 1980s. In 1988, scientist James Hansen gave the U.S. Congress a strong warning. Earth's temperatures were rising, and humans were the cause of this global warming.

At the time, many people did not think global warming was a problem. Some scientists disagreed with Hansen. Some of these scientists had the backing of the fossil fuel industries, which release greenhouse gases into the environment. These industries did not want to admit their part in climate change. Changing their practices to reduce greenhouse gas emissions would cost them a lot of money.

James Hansen is a former NASA scientist turned climate activist. He has continued his activism in the decades since his speech to Congress. He participates in protests, has written books, and gives speeches.

Key Events

Earth Summit

The first Earth Summit took place in Rio de Janeiro, Brazil, in 1992. The event gathered 117 world leaders and people from 178 countries. One of the agreements made the United Nations Framework Convention on Climate Change. Countries agreed to control their greenhouse gas emissions. It paved the way for the 2016 Paris Agreement.

International Cooperation

An international agreement, called the Kyoto Protocol, tried to address the problem. It called for industrialized countries to reduce greenhouse gases. Many countries signed, including the United Kingdom, France, and Australia. The United States refused to follow the guidelines. Canada signed the protocol, but pulled out at the second phase in 2012.

In 2015, the United Nations brought together world leaders to sign the Paris Agreement. This aimed to lower greenhouse gas emissions and keep down global warming. It set a target to keep global temperature rise below 2 degrees Celsius (3.6 °F) and below 1.5 degrees Celsius (2.7 °F) if possible.

Hansen (in the brown hat, bottom right) was arrested with about 100 other people outside the White House in September, 2010, at a protest against mining for coal.

CHAPTER 6

MAKING A DIFFERENCE TODAY

Young activists have become leading voices in the fight to protect the planet. They have become organized. By putting their voices together, young activists are making governments and world leaders listen.

Youth-led activist groups Zero Hour and the Sunrise Movement formed in 2017. The U.S. Climate Strike and FridaysForFuture formed the next year. These youth groups started small but grew quickly around the world. They used social media to spread their message. They are demanding that leaders see climate change as an emergency. They want governments to stop the use of fossil fuels and to support renewable energy sources.

Young activists from around the world marched in the Global Climate Strike on September 20, 2019.

Different Actions

To force changes, young activists have taken many actions. These include protesting in front of the government offices in Washington, DC, and worldwide school strikes for climate. In India in 2017, nine-year-old Ridhima Pandey took legal action against the Indian government for failing to act on climate change. Many of these young leaders are being heard. In 2019, the UN held the first Climate Youth Summit in New York. The event brought together young activists, inventors, and other young leaders together with world leaders. It was a chance for them to share their thoughts and their ideas to fight climate change. The UN also involves young environmentalists in their Young Champions of the Earth program. The program awards and supports people who have ideas for change, such as new inventions.

Key Voices

Greta Thunberg

In 2018, Greta Thunberg began sitting in front of the Swedish government building every Friday. She was protesting the government's lack of action on climate change. #FridaysForFuture went viral, and other youth groups became involved. Friday school strikes were staged around the world. More than 4 million people joined the Global Climate Strike.

Keep It in the Ground

The use of fossil fuels is an ongoing issue. Fossil fuels are not renewable. They release greenhouse gases and cause pollution. Climate activists have asked the world to stop using and putting money into fossil fuels.

The Paris Agreement (see page 35) does not say that countries need to limit the production of fossil fuel. In 2016, groups such as 350 did not think the agreement went far enough. They organized a worldwide protest called Break Free. Thousands of people around the world took part in protests. They called for fossils fuels to stay in the ground and for a complete move to renewable energy sources.

Some of the largest protests took place in the Philippines. In 2016, around 10,000 people marched against a new coal-fired power plant in the city of Batangas.

Plastic Pollution

Plastic is a **by-product** of the fossil fuel industry. Plastic waste has become an urgent problem. We make more than 300 million tons (272 million metric tons) of plastic each year, half of which is single-use plastic. Only 9 percent of plastic is recycled. The rest ends up in landfill and in oceans. Plastic does not break down. It often separates into tiny pieces, called microplastics. In oceans, these can kill marine life that eat it by mistake. The Great Pacific Garbage Patch is an area of the Pacific Ocean that contains about 1.8 trillion pieces of plastic.

The Great Pacific Garbage Patch, in the Pacific Ocean, covers an area about three times the size of France, or more than twice the size of Texas!

Key Voices

Boyan Slat

Dutch inventor Boyan Slat was a 16-year-old student when he went scuba diving in Greece. He was shocked by the amount of plastic he saw in the sea. Slat devoted a high school project to ocean plastic pollution. He founded his organization OceanCleanUp when he was 18. The organization's scientists and engineers are designing systems to clean up the plastic that is already in the oceans and to stop other plastic from reaching them.

Looking Forward

More than ever, people around the world are worried about the environment. Although people have been trying to protect the planet for centuries, the work is not done.

Many politicians have become environmental activists. In the United States, Alexandria Ocasio-Cortez and Ed Markey led the fight for the Green New Deal. This called on the government to reduce carbon emissions, create new jobs, and provide training, especially in communities that rely on work in the fossil fuel industry. In Europe, the European Union voted to support their own Green Deal in 2020. It set out a plan to make Europe **climate neutral** by 2050.

Alexandria Ocasio-Cortez talks about the importance of the Green New Deal. Although the government did not approve the deal, the deal sparked talk about climate change.

GOOD JOBS AND A LIVEABLE FUTURE

Collective Action

Today's activism is worldwide and powerful. In 2015, the United Nations published 17 Sustainable Development Goals. Climate action and environmental protection are key parts of the initiative, which encourages youth to get involved in creating a sustainable future. Greta Thunberg was *Time* magazine's Person of the Year in 2019. The youth movement has sparked a growing generation of activism.

Nadia Nazar is a co-founder of Zero Hour, an organization that supports and organizes youth activism for the environment. She spoke at the UN's International Day of the Girl summit in 2018 about the impact climate change has on girls.

Key Voices

Extinction Rebellion

Extinction Rebellion formed in 2018. The group organizes nonviolent acts, such as blocking traffic, to draw attention to the climate change crisis. In April 2019, activists staged a sit-in protest in Central London, England. The group demands that the government take action to stop climate change and protect the environment.

GET INVOLVED

Anyone can become an activist to protect the planet. Here are some suggestions of how to contribute.

1 Start Local

Environmental activism can start at home! As a family, recycle as much as possible and cut out single-use plastics (such as plastic water bottles, takeout cups, and plastic bags). Get involved in local park or beach clean-ups. Start a campaign to get your school or neighborhood businesses to stop using single-use plastics.

2 Consumer Power

When you shop, look for zero waste and eco-friendly products. Arrange a clothing exchange with friends rather than buying new. Look for products with minimal plastic packaging. Zero-waste stores are starting up, where you can take an old bottle and refill it with toiletries or cleaning materials.

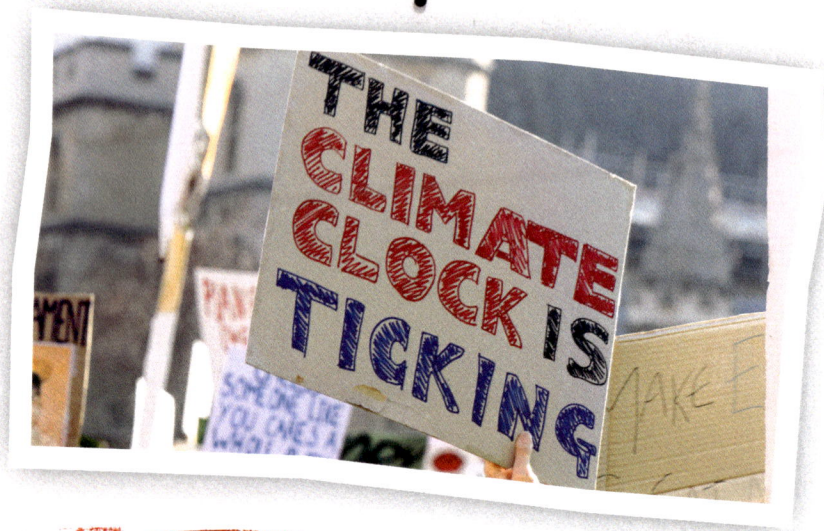

3 Volunteer

Environmental organizations need volunteers to help with a range of activities, from raising money to helping in the office. There are charities that focus on climate change, plastic pollution, protecting wildlife, and more. Volunteering is a great way to learn about a subject, as well as meet like-minded people.

4 March!

Watch for events in your area. If there's a march, make a sign, tell others about the event, show up, and display your support.

5 Join Up!

Most organizations have an e-mail list you can join for free. Sign up and stay in touch with what's happening in the fight for the planet!

Timeline

1306 King Edward I bans the burning of sea coal in England.

1890 Yosemite becomes a national park.

1892 Sierra Club is founded.

1955 Air Pollution Control Act is passed in the United States.

1956 Clean Air Act is passed in the United Kingdom.

1961 The World Wildlife Fund is started.

1962 Rachel Carson's book *Silent Spring* highlights the damage caused by chemical pesticides.

1970 The First Earth Day takes place on April 22; it becomes an annual event.

1971 Greenpeace is founded.

1977 Wangari Maathai creates the Green Belt Movement in Kenya; she is awarded a Nobel Prize in 2004.

1987 The Montreal Protocol bans the chemicals that had created the hole in the ozone layer.

1988 Scientist James Hansen gives evidence to the U.S. Congress about global warming.

1992 The First Earth Summit takes place in Brazil.

2012 The Idle No More movement is started in Canada.

2016 The UN Paris Agreement sets targets to reduce greenhouse gas emissions.

2016 Break Free protests take place around the world; thousands of people, led by Indigenous peoples, protest the Dakota Access Pipeline.

2017 Youth-led environment movements Zero Hour and the Sunrise Movement are established.

2018 Extinction Rebellion begins a series of nonviolent direct action; Greta Thunberg starts a school strike for climate change that grows into #FridaysForFuture.

2019 The UN holds a Climate Youth Summit in New York; Greta Thunberg is named *Time* magazine's Person of the Year.

Sources

Chapter 1

"Can the Earth Be Saved?" The Nature Conservancy. December 16, 2019. https://bit.ly/2DjQvJT

"Saving Earth." *Encyclopedia Britannica.* www.britannica.com/explore/savingearth

Weyler, Rex. "A Brief History of Environmentalism." Greenpeace. January 5, 2018. https://bit.ly/2PteSHk

Chapter 2

Adam, David. "Earthshakers: the top 100 green campaigners of all time." *The Guardian.* November 28, 2006. https://bit.ly/2ES66R4

Suzuki, David. "Aboriginal people, not environmentalists, are our best bet for protecting the planet." *Vancouver Sun.* June 8, 2015. https://bit.ly/3aBLhW6

Chapter 3

"12 historic American environmentalists who made our wilderness all-star draft." The Wilderness Society. September 11, 2013. https://bit.ly/3gBKGpf

Wilkinson, Katharine. "The Woman Who Discovered the Cause of Global Warming Was Long Overlooked. Her Story Is a Reminder to Champion All Women Leading on Climate." *Time.* July 17, 2019. time.com/5626806/ eunice-foote-women-climate-science

Chapter 4

"How Does Oil Get into the Ocean?" NOAA Office of Response and Restoration. November 2, 2015. www.response.restoration.noaa.gov/ about/media/how-does-oil-get-ocean.html

Thulin, Lila. "How an Oil Spill 50 Years Ago Inspired the First Earth Day." *Smithsonian Magazine.* April 22, 2019. https://bit.ly/3gBaPEJ

Chapter 5

Elbein, Saul. "These Are the Defiant 'Water Protectors' of Standing Rock." *National Geographic News,* January 26, 2017. https://on.natgeo.com/3fFxVJ8

Milman, Oliver. "Ex-Nasa scientist: 30 years on, world is failing 'miserably' to address climate change." *The Guardian.* June 19, 2018. https://bit.ly/2F3AoRf

Chapter 6

Kaur, Harmeet. "7 startling facts about the crisis facing our oceans." *CNN.* June 8, 2019. https://cnn.it/3h9wuEq

Worland, Justin. "How the Green New Deal is Forcing Politicians to Finally Address Climate Change." *Time.* March 21, 2019. time.com/5555721/green-new-deal-climate-change

Glossary

acid rain Rain, snow, or fog that contains high levels of acids, which kills plants and fish

acts Laws made in government

arable Suitable for farming

atmosphere Layers of gases that surround and protect Earth

biodiversity Variety of plants and animals found in a particular habitat or on Earth

boycott A punishment by which people refuse to buy from a company or store as a form of protest

by-product Something that is made as a result of the production of something else

climate change Change in Earth's climate over a long time

climate neutral Having a balance between the amount of greenhouse gases added to the atmosphere and the amount removed from the atmosphere

conservation Preservation, protection, or restoration of environments and species

contamination Something harmful added, such as a poison

deforestation Removing forest from land that is then used for another purpose

desertification The process in which an area of land becomes desert

discredit To harm the good reputation of someone or something

ecosystems Communities of living and non-living things in an environment

eroded Gradually wore down or destroyed

extract To remove or dig out of the ground

food chains Chains of organisms in which each member uses the member below as food

fossil fuels Fuels such as oil and coal that form from animals and plants that died long ago

fracking Method of getting oil or gas from rocks by injecting high-pressure fluids

global warming The gradual increase in Earth's overall temperature, caused by human activity

grassroots Describes groups of people who are not in positions of power

greenhouse gas A gas such as carbon dioxide that traps heat

habitable Suitable for living

Indigenous Native to, or having always lived in, a specific place

industry Businesses that process raw materials or make things in factories

logging The business of cutting down trees

native Naturally occurring in a specific place

natural resources Raw materials found in or on the ground, such as wood

non-renewable Not able to be made again

nuclear testing Testing nuclear bombs by exploding them above or below the ground

nutrients Substances, such as vitamins, that living things need to grow

ozone Layer of Earth's atmosphere which absorbs most of the ultraviolet radiation coming from the Sun to Earth

pesticides Chemicals used to kill insects or animals that damage crops

pollution Introduction of harmful substances into the environment

refinery A facility in which a substance is refined, or changed

reservation A designated area of land given to Native American tribes and First Nation bands by the American and Canadian governments

sea coal Coal that washes up on beaches

smog Dirty air caused by pollution

sustainable Keeping an ecological balance to avoid overusing natural resources

tactic Action or method

toxic Poisonous

transcontinental Across a continent

Further Information

Books

Berners-Lee, Mike. *There Is No Planet B: A Handbook for the Make or Break Years.* New York: Cambridge University Press, 2019.

Croy, Anita. *Rachel Carson* (Scientists Who Changed the World). New York: Crabtree Publishing Company, 2020.

Margolin, Jamie. *Youth to Power: Your Voice and How to Use It.* New York: Hachette Books, 2020.

Sjonger, Rebecca. *Taking Action to Help the Environment* (UN Sustainable Development Goals). New York: Crabtree Publishing Company, 2020.

Stefoff, Rebecca. *The Environmental Movement: Then and Now* (America: 50 Years of Change). North Mankato, MN: Capstone Press, 2018.

Thunberg, Greta. *No One Is Too Small to Make a Difference.* New York: Penguin Books, 2019.

Websites

www.undp.org/content/undp/en/home/sustainable-development-goals.html
Learn more about the UN Sustainable Development Goals. Goals 13, 14, and 15 focus on protecting the environment.

www.earthday.org/our-work
Information about Earth Day and its global campaigns.

www.endangered.org/10-easy-things-you-can-do-to-save-endangered-species
Learn about endangered species and how to protect them.

www.nationalgeographic.com/environment/climate-change
Information about climate change from National Geographic.

oceana.org/our-campaigns/plastics
Find out how to fight plastic pollution of the oceans.

www.unenvironment.org/explore-topics/climate-change/facts-about-climate-emergency
Facts about the climate emergency from the United Nations environment program.

Index